1 ONE-YEAR OLD

JULIET BAWDEN with photographs by HELEN PASK

Henry Holt and Company

1 one-year-old

2 two-year-olds

3 three-year-olds

4 four-year-olds

5 five-year-olds

6 six-year-olds

7 seven-year-olds

8 eight-year-olds

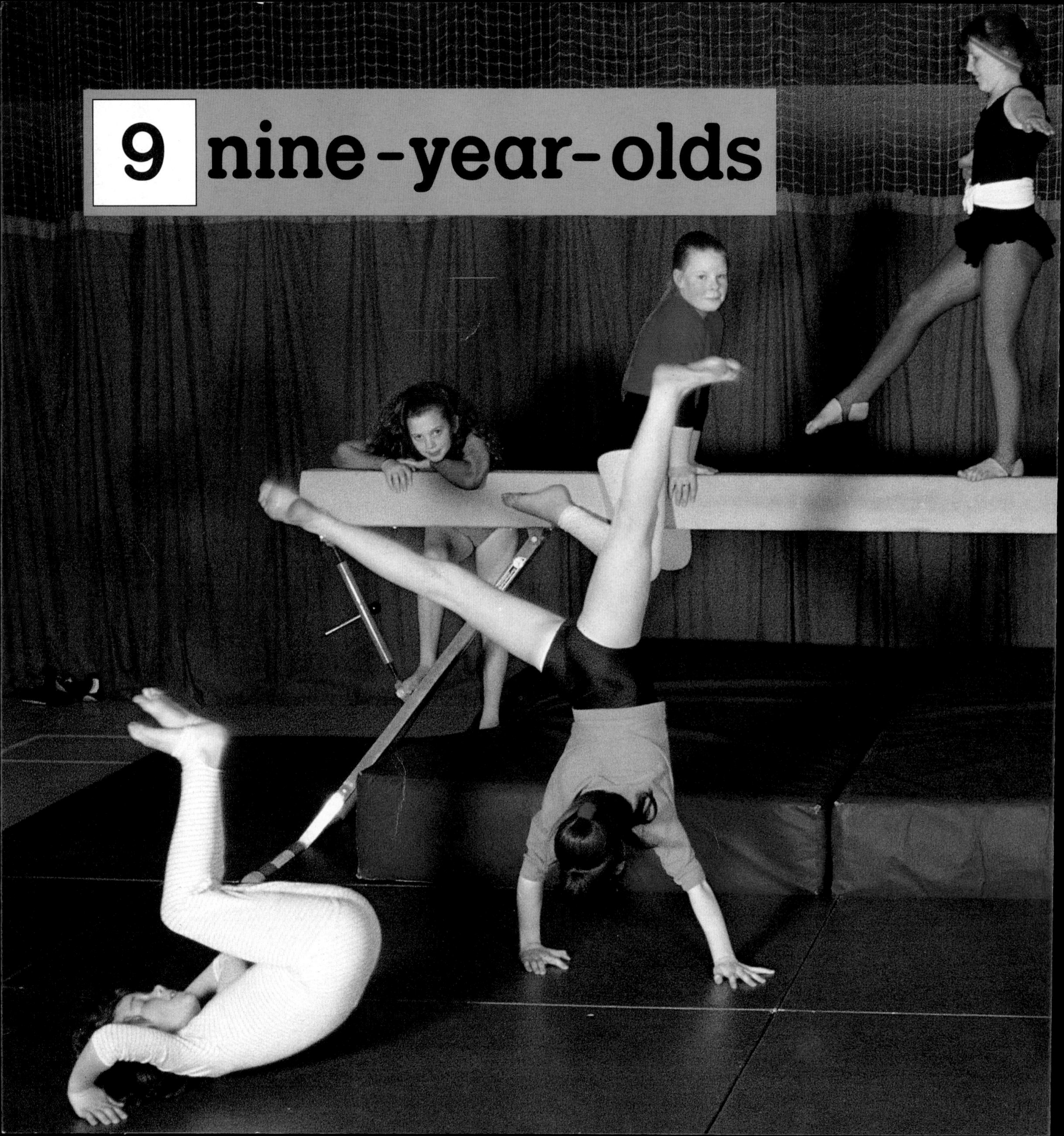

9 nine-year-olds

10 ten-year-olds